Issue 2, August 2019

©Stir Research Technologies, India

Stir Research Technologies is a Research and Development firm which works on collaborative research projects based on Artificial Intelligence and Friction Stir Welding. It is the first firm to develop MOOCs on Friction Stir Welding process.

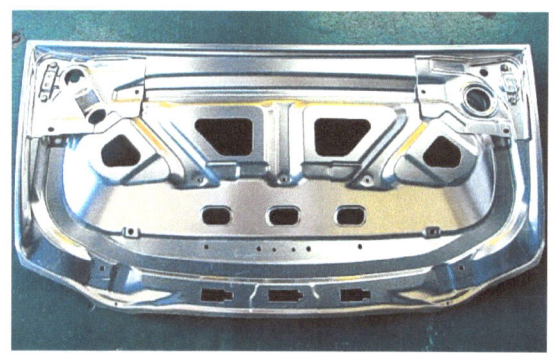

The friction Stir Welding (FSW) process was invented by Wayne Thomas of the Welding Institute in 1991, at Cambridge, in United Kingdom. TWI has further developed this process and patented it.

Friction stir welding (FSW) is a solid-state joining process. Solid state welding means that there is no molten state included in joining or welding the workpiece. This joining technique saves energy and is eco friendly. It is mostly used to weld aluminium materials in automobile and aerospace industries.

The FSW is the process in which a tool is used which does not gets consumed in joining the workpiece so it is called as non-consumable tool. A non-consumable rotating tool with a pin and a shoulder is inserted into the adjacent edges of sheets or plates to be joined and moved along the line of joint till the end .It is considered to be the most significant development in metal joining process and is a "green" technology due to its energy efficiency, environment friendliness etc.

This issue will focus on the application of Friction Stir Welding process in automobile industries.

Honda developed Friction Stir Welding technology to weld steel and aluminum together

Although some engineers have had success in spot welding steel and aluminum together, it has largely been considered impossible to achieve reliable, continuous welds directly between the two dissimilar metals. The results, which include lower vehicle weight and better performance, can be seen in the 2013 Accord.

A diagram of an Accord subframe made using the new welding process

The Honda team developed a variation on Friction Stir Welding, in which metals are joined via mechanical pressure – it's the same technique that has been used for experimental steel/aluminum spot welds in the past. As Honda explains it, "This technology generates a new and stable metallic bonding between steel and aluminum by moving a rotating tool on the top of the aluminum which is lapped over the steel with high pressure." The welds that result are reportedly as strong or stronger than those made using regular Metal Inert Gas welding.

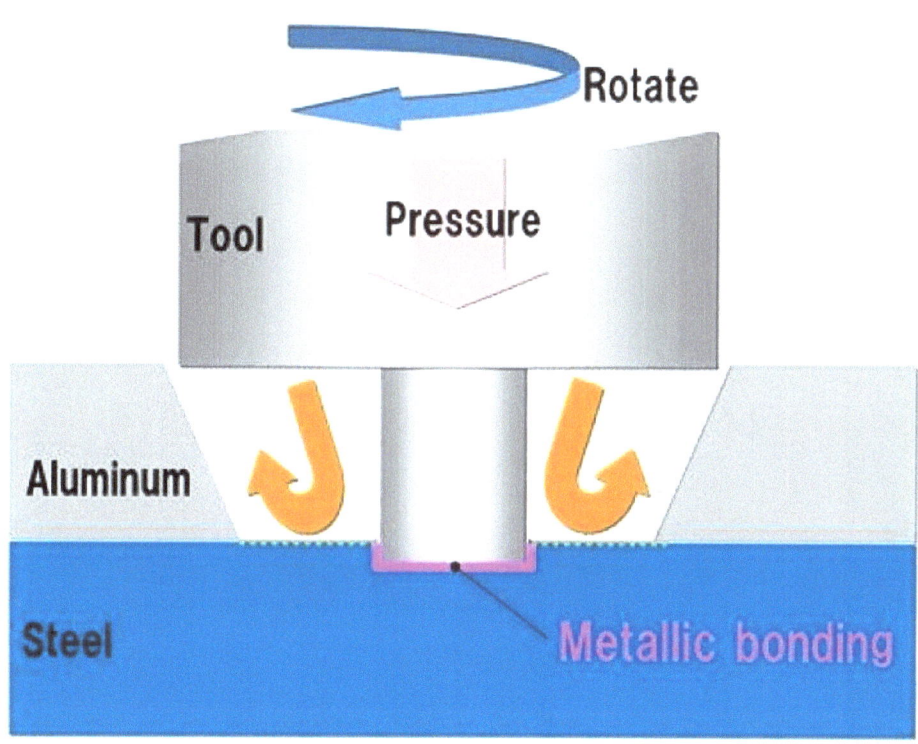

A diagram illustrating Honda's variation on Friction Stir Welding

Steel/aluminum subframes built with the new technique are said to be 25 percent lighter than those made entirely out of steel, which should translate into improved fuel economy. The process also made it possible to alter the structure of the subframe, so that the mounting point for the suspension could be relocated – this change has reportedly increased the rigidity of the mounting point by 20 percent, and thereby improved the car's dynamic performance.

Additionally, the new process uses about half the amount of electricity as Metal Inert Gas welding, and the machinery it requires isn't as large as that traditionally used for Friction Stir Welding – in fact, it can be attached to an industrial robot. The technique can also be used for aluminum-to-aluminum welding, without any hardware changes.

A new non-destructive inspection system, incorporating an infra-red camera and a laser, is used to check all of the steel/aluminum welds.

Source: **Honda**

Robotic Friction Stir Welding for Automotive applications

Since the invention of friction stir welding (further abbreviated as FSW) 18 years ago, it was obvious that this technique had some significant advantages regarding automotive and aviation industries. However, a breakthrough is never achieved so far. This is partially because FSW has to compete with some very common and reliable techniques like arc and spot welding. Also Saab uses spot welding very often. In general, a regular car contains between 2000 and 3000 spot welds. Due to the introduction of robotic FSW, the process gains in flexibility which makes it now attractive to explore the possibilities and limits for automotive applications. If it turns out that FSW is as fast and as flexible as spot welding, it can be considered to replace the spot welding installation by friction stir welding robots.

Volvo Aero has also economical reasons to consider friction stir welding. For their production of engine parts, they often have to purchase very large parts in titanium and inconel with diameters up to 3.5m. Since only a few companies are able to fabricate those large parts, this has always been an expensive factor in the production chain. Friction stir welding allows Volvo Aero to produce these large parts itself. It will also allow smaller manufactures to produce these big parts, what might cause a price drop for these products.

Although this new welding technique still needs some further development for high strength materials, an own production is very realistic and can result into a more cost efficient process.

Researchers at University West in Trollhättan have addressed two drawbacks to the robotic friction stir welding (FSW) joining process for mixed materials—path accuracy and temperature—with the development of a deflection model and integral temperature controller.

Car manufacturers are increasingly looking to a hybrid or mixed materials design, in which a combination of different materials such as steel and aluminium are joined, for weight reductions in their vehicles. With classic welding methods, joining of dissimilar materials has not been possible. With friction stir welding (FSW), on the other hand, high quality dissimilar joints can be obtained. The use of industrial robots also allows FSW of materials along complex joint lines.

In FSW, a rotating non-consumable cylinder is pressed into the material. The combination of frictional heat and the mechanical "stirring" creates a high-quality welding joint, without melting the material. The welding temperature is kept below the melting point, which means that the alloy properties are not destroyed and strong joints are achieved. There have been two practical challenges with this method:

- Robot compliance. This results in vibrations and insufficient path accuracy. For FSW, path accuracy is important as it can cause the welding tool to miss the joint line and thereby cause welding defects.
- Variable heat dissipation on complex geometries. Variable heat dissipation in the workpiece causes great variations in the

welding temperature. Especially for force-controlled robots, this can lead to severe welding defects, fixture- and machine damage when the material overheats.

To address the first issue, the researchers first measured path deviations post-weld and later by using a camera and laser distance sensor to measure deviations online. Based on that knowledge, they created a robot deflection model. The model is able to estimate the tool offset during welding, based on the location and measured tool forces. This model can be used for online path compensation, improving path accuracy and reduced welding defects.

To address the second issue, they developed a new temperature method which measures the temperature at the interface of the tool and the workpiece, based on the thermo-electric effect. The temperature information is used as input to a closed-loop temperature controller. This modifies primarily the rotational speed of the tool and secondarily the axial force.

The controller is able to maintain a stable welding temperature and thereby improve the quality and allow joining of geometries which were impossible to weld without temperature control.

The resulting robot welds with higher precision and with the temperature controller it only takes a few hours to program 3D joints, said Dr. Jeroen De Backer, who wrote his thesis on this new method. Manual programming of a similar component took up to a week.

With the aid of the robot, and the temperature measurement, the researchers have also been able to weld advanced three-dimensional joints. This enables the welding of small and more complex components with curved surfaces. Furthermore, the energy consumption of FSW is lower than when using conventional welding methods.

Learn Friction Stir Welding via online MOOCs

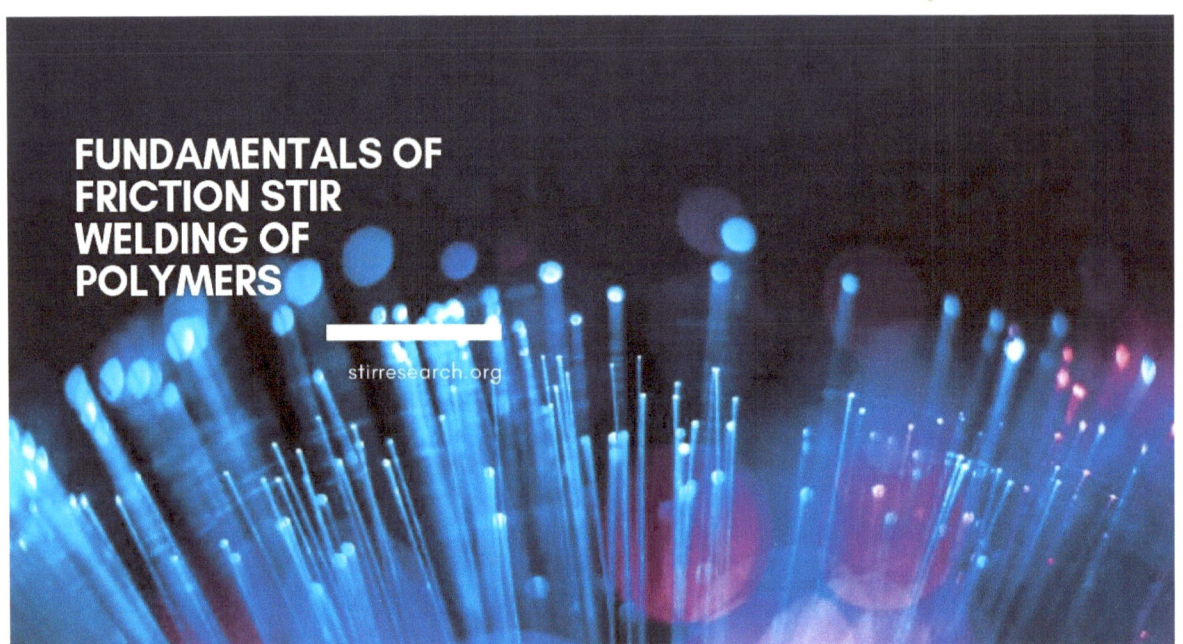

Available on Udemy

Friction Stir Welding Now Suitable for High-Volume Auto Assembly

Engineers from the Pacific Northwest National Laboratory (PNNL) have improved friction stir welding (FSW) so that it can now join aluminum sheets of varying thicknesses, a key capability for making welds that are strong in places where needed, yet light overall. They've also transformed FSW so it is 10 times faster than conventional FSW, which is limited to welding at about 20 in./min. The higher speeds mean FSW can meet the auto industry's high-volume assembly requirements.

In FSW, a pin tool spins against the edges of two pieces of metal positioned next to each other. As the pin travels along, it creates friction that heats, mixes, and joins the alloys without melting them. To optimize the process, researchers created pins of different shape, length, and diameter, and used them in FSW but varied the depth, rotational speed, and toll angle on the pins. They then used statistical analysis to identify the most productive combination of tool and weld parameters that could support high-speed production. They also discovered that the faster FSW is carried out, the stronger the resulting welds.

In a test, the updated FSW process was used to join two sheets of aluminum, one much thicker than the other. The joined sheet was then stamped into a car door. This FSW was quicker and less expensive than conventional manufacturing methods and turned out a door that was 62% lighter and costs 25% less.

PNNL researchers are working on this project with General Motors, Alcoa, and TWB Co., with two years of funding remaining. They will use that time to get faster weld speeds and refine the process so it can be used around the contours and corners of complex aluminum parts, for which laser welding is not commercially feasible. The team also is modifying FSW to join different alloys, such as automotive-grade aluminum alloys with light, ultra-high strength alloys currently used only in aerospace applications.

Source: Machinedesign.com

Friction Stir Welding & Electric Cars

Sales figures for battery-driven electric cars ((B)EV) and hybrid vehicles (HEV) are rapidly increasing, since they are becoming more economical and less expensive due to the further development of key components such as energy storage systems. For automotive manufacturers this opens up a market potential which can be successfully exploited, provided that they are able to counteract unique features and limitations by appropriate adaptations of vehicle concepts and the manufacturing process in an economically viable way.

A hybrid drive is used in vehicles with internal combustion engines to improve efficiency and reduce fuel consumption. Numerous variations are manufactured for series-production vehicles. Only the variant "plug-in hybrid vehicle (PHEV)", in which the energy storage device can also be charged from a mains power supply, will be considered here. These vehicles are equipped with energy storage systems which allow longer distances, ranging from 60 to 80 km, to be travelled using the electric drive only. All-electric vehicles are characterised by the fact that they are heavier compared to vehicles with internal combustion engines and the energy storage device constitutes a substantial part of the weight and also of the costs involved.

Although electric vehicles demonstrate a higher degree of efficiency in the range between approx. 65 and 80% and some heavy components such as internal combustion engine, transmission and exhaust system can be dispensed with, the energy density is lower by a factor of 50 to 100 when compared to petrol-fuelled engines. This results in a large volume and a high weight, which must be safely accommodated in the vehicle. For a range of 400km commercially available energy storage devices have a weight of approx. 320kg.

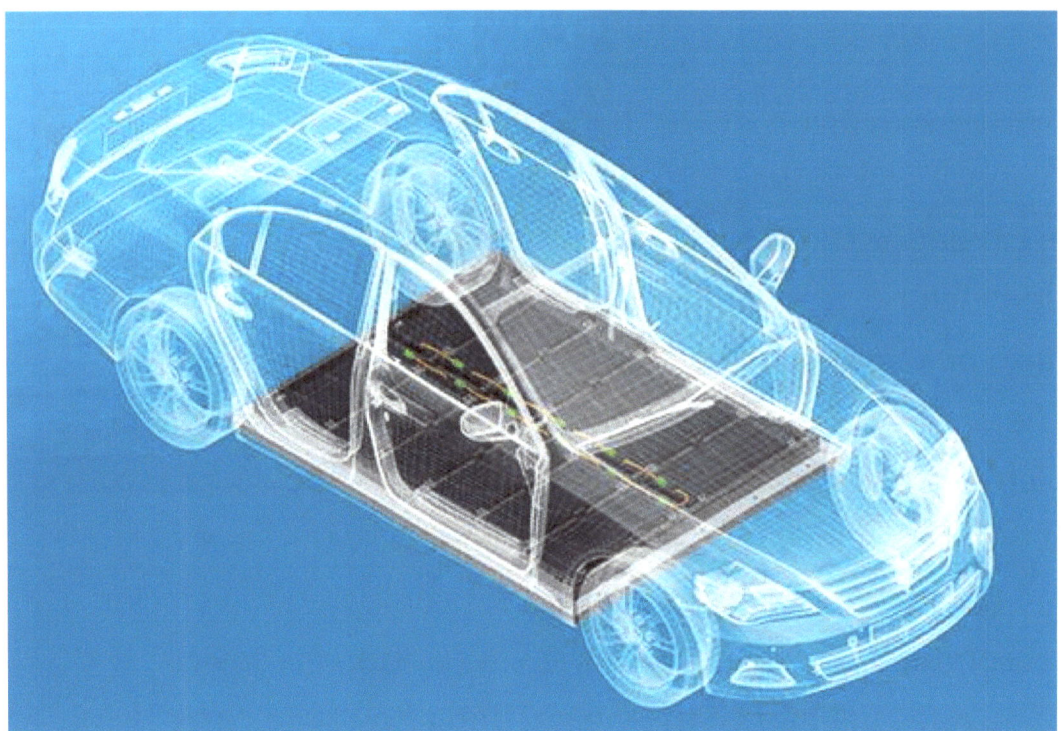

What specific demands on the joining technology, with regard to vehicle concepts and the manufacture of components, do the unique features described above create? In this context, passenger cars have primarily been addressed; it is particularly difficult to develop a vehicle concept for them when compared to buses and commercial vehicles. Bulky and heavy battery systems must be in- stalled in plug-in hybrid and electric vehicles in such a way that they have a low center of gravity and restrict the passenger room and luggage compartment as little as possible. These systems must be securely mounted and be able bear the vehicle-specific loads over their life span. Furthermore, the systems must not impair the crash behavior and must take into account additional hazards, such as fire and environmental pollution in dangerous situations.

In order to guarantee temperature control, some manufacturers integrate cooling systems into the floor on which battery modules with a low thermal transition resistance are mounted. The difficulty in the manufacture of the storage housing lies in having to produce a closed, pressure-tight cooling circuit, which withstands the stress caused by the alternating pressure, causes low thermal deformation of the mounting surface and allows a permanent bonding of different aluminium alloys. Here, the friction stir welding process has proven to be an optimum joining technology. The cooling circuit is closed by welding a base plate to the cast housing. For this purpose, an aluminium sheet of alloy EN AW 6000 has to be welded to a cast body of the EN AC 4000 class over a length of more than 10 m. Depending on the part design, the weld seams are executed in a meandering shape or as overlapping joints. The demands on weld seam quality are very high, in particular due to the requirements of the alternating pressure test with more than 200,000 pressure changes. In addition, the flatness of the mounting surfaces inside the housing should be less than 0.2 mm above that of the mounting surface, even after welding.

The battery housings for plug-in hybrid vehicles are usually produced as a die-cast component. If cooling circuits are to be integrated directly into the housing, this area must be subsequently closed by welding on an aluminum sheet. For this purpose, for example, a sheet made of an EN-AC 6000 alloy, 3 mm thick, must be welded as an overlapping joint to the cast body, an EN-AC 4000 alloy. The welding area is usually milled during machining of the cast body, so that the cast skin and irregularities in the con- tact surface pose no problem. The difficulty lies in designing the component in such a way that the minimum cross-section in the welding region of the cast body is sufficient to ensure ample rigidity also in the area of the cooling water connections. For good heat transfer, the flatness of the mounting surfaces for the modules after welding is decisive for optimum cooling of the battery pack. For this reason, the housings in these areas must be supported in order to reduce distortion. Weak points in components under pressure are the process-related exit holes at the end of each weld, caused by the tilt angle and the displaced material. Seam ends should therefore always be located outside the area under pressure. If this is not possible, or if additional weld seams need to be set for minimizing the surface under pressure, the end of the seam must be welded circumferentially by "closing in" the probe's exit hole.

Where all-electric vehicles are concerned, the storage housings consist of extruded sections, welded together to create a frame structure by arc welding. Following machining of the seams in the welding area of the base, these frames are welded to a base sheet or base plate, consisting of several double-walled extruded sections, to form a liquid-tight housing by means of the friction stir welding process.

The friction stir welding process is a solid phase joining process in which the materials to be joined are not melted to form a weld pool. Rather, the parts being joined are "stirred" by means of a rotating tool in the plastic state below the solidus. A tool, consisting of a shoulder and a probe, are generally used to carry out friction stir welding. The welding probe is positioned centrally under the shoulder. The parts being joined are clamped firmly to the welding fixture. In the convention- al process, shoulder and probe are rotated about their own axis and are pushed into the joint line between the two workpieces with a defined force. In addition, there are processes utilizing a fixed shoulder and rotating probe. Frictional heat is generated between the tool and the parts to be joined causing the material to plasticize. Thanks to the special design of the welding pin, the work- pieces are joined before the softened material in the fusion zone solidifies again.

When using a double-walled base plate, this must be welded from both sides from the individual extruded sections in a preliminary operation. In doing so, the geometric tolerances of the extruded precision profiles, as defined in Standard EN 12020-2, are to be observed. If welding of the profiles on the longitudinal edges is impossible due to the standard tolerances, then these profiles must be machined prior to welding. For example, when welding double-walled base plates of 12 mm thickness, a gap of 0.3 mm can be reliably welded at a penetration depth of approx. 4 mm using the stationary shouldered tool without a compaction defect occurring. When using a classic tool with a welding pin, 5 mm in diameter, and a shoulder with a diameter of 13 mm, the gap cannot be more than 0.5 mm at the most. The gap volume has a considerable influence on the interface at the bottom of the seam and must be taken into consideration when designing the components.

It would not be practical to weld the base plate down through the full material thickness of 12 to 15 mm, since a very wide seam with a shoulder diameter in excess of 20 mm is required. The down forces necessary and the bonding cross-section of more than 10 mm cannot be achieved in the frame section since this is not compatible with the objective of a lightweight construction. Consequently, the base plate is mechanically machined in the bonding area and the hollow-chamber profiles welded to an external bar. Where cooling channels are integrated in the base plate, further mechanical processing steps are needed to be able to weld the profiles in a reliable and pressure-tight manner. A more widespread concept for energy storage housings for all-electric vehicles is a frame section with additional struts which is sealed by means of a base plate. The friction stir welding process is employed to weld the base plate around the periphery such that it is sealed against liquids. The struts are also welded to the base plate, to absorb the weight of the battery: either MIG seams are welded after the base plate has been joined or friction stir welding is used to weld through the base plate into the struts to form an overlap joint.

For this latter option attention must be paid that the seam end with the exit hole lies outside the leak tightness area. A process-compatible design of the frame profiles in the welding area must be ensured for both the base plate and bottom section concepts. The stiffness of the profiles is one of a major factor, which is explained in more detail in the following. The frame sections with the side members are part of the vehicle's structure for most manufacturers and must be designed to meet the requirements for side crashes. In addition, the forces generated during the friction stir welding process must be taken into consideration for these frames and also for the trans- verse profiles. In order to achieve traversing speeds of around 1m/min, thereby enabling the required heat input into the joint line, a sufficiently high friction must be created. To do so, the friction stir weld tool must be pressed onto the surface of the components to be welded with a force ranging between 3.5 and 5 kN. This down force must be absorbed by all types of profiles in the welding area and by the clamping fixture without significant distortion. The wall thickness in the welding area should also be defined to suit the process.

It should be borne in mind that the material surrounding the welding pin becomes plasticized and therefore cannot contribute to rigidity and stiffness of the structure. If the welding pin is long enough to penetrate the profiled section by approx. 0.2 to 0.5 mm, then it must be assumed that the plasticised area, the thermo-mechanically affected zone, has a depth of approx. 0.4 to 0.8 mm. In practice, wall thicknesses of approx. 3.5 mm have been tried and proven where welding in the area of a profile edge is possible and represent a compromise between lightweight construction and stiffness when welding with high process forces. The beginnings and ends of the seams can cause distortion of the pro- files at this wall thickness. These deformations can also result in superficial compaction defects. With profile thicknesses of 4.0 mm in the welding area and using 3 mm thick base plates, a reliable, safe seam without distortion can also be achieved in other profile areas.

If one now considers the demands made by the automotive industry on the joining technology for the e-mobility sector and the advantages of the friction stir welding process for some of the main joining applications, then considerable market potential for production lines incorporating the friction stir welding process can be seen over the coming years. Along with the technological equipment needed to make use of the process as a safe and economical production method, however, the demands for transparency in the collection of quality data and trace- ability of the key process parameters must be fulfilled, with reference to each individual vehicle. With this in mind, KUKA has formed the basis for the integration of Industry 4.0 with their Process Control and Documentation sys- tem (PCD), which enhance our customers' pro- duction facilities.

Source: http://dx.doi.org/10.17729/ebis.2017.5/11

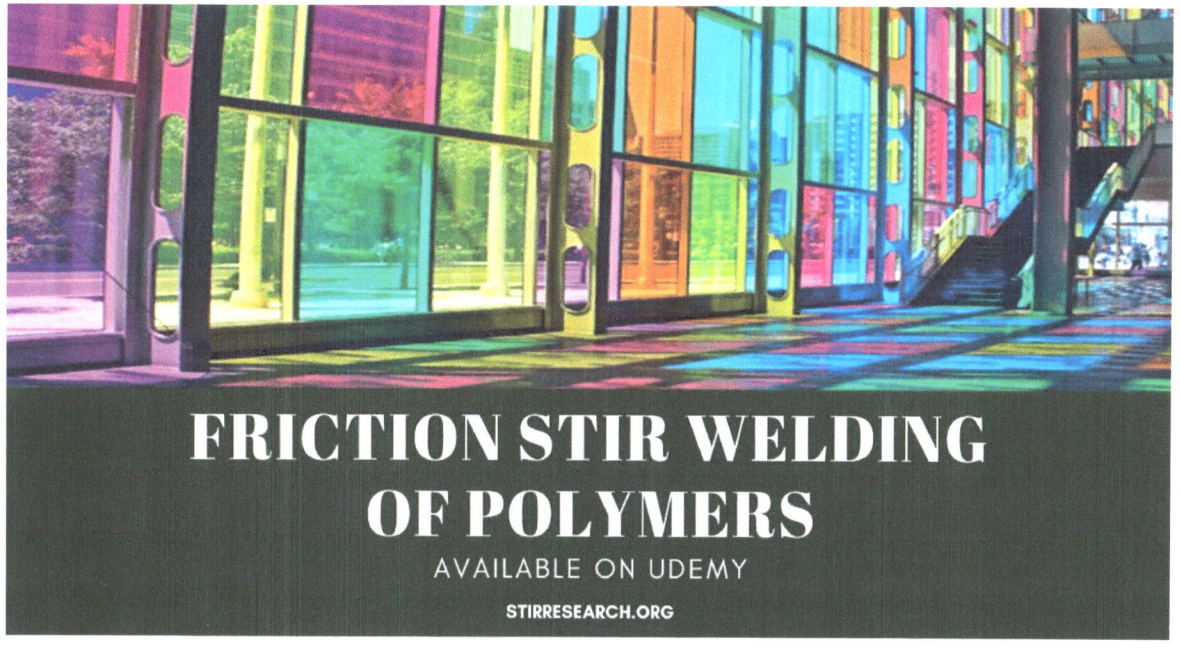

Plastics have become candidates for an increasing number of applications, it is important to continue developing processes to join them efficiently. In the past, a few processes have dominated the area. Now a new process, Friction Stir Welding, enters the arena of competition with the established practices.

When compared on these bases, Friction Stir Welding stands as an able competitor with established processes.

Akshansh Mishra, Founder and Project Scientific Officer of Stir Research Technologies developed the 5th MOOC of Friction Stir Welding series entitled FRICTION STIR WELDING OF POLYMERS which is available on Udemy. We assure you that you are surely going to enjoy and explore new things in this course.

This course deals with the modern joining method so called Friction Stir Welding available on Udemy. In this course, the learners will gain knowledge of the fundamentals of Friction Stir Welding process, importance of tool design and fabrication of fixture for Friction Stir Welding process. Assignments are given to the learners for developing their understandings in the field of applications of Friction Stir Welding process. This is an introductory course and continuation of this course entitled Advance Study in Friction Stir Welding process is under development. As an attachment link to the research papers and study materials are provided.

Fundamentals of Friction Stir Welding

This course outlines the basics topics like tool design, material flow mechanism involved in Friction Stir Welding process. This course is being featured on Udemy and Skillshare.

www.ingramcontent.com/pod-product-compliance
Lightning Source LLC
Chambersburg PA
CBHW051838210526
45473CB00005B/1929